Thomas Wharton Jones

Evolution of the Human Race

from apes, and of apes from lower animals. A doctrine unsanctioned by

science.

Thomas Wharton Jones

Evolution of the Human Race
from apes, and of apes from lower animals. A doctrine unsanctioned by science.

ISBN/EAN: 9783337240912

Printed in Europe, USA, Canada, Australia, Japan

Cover: Foto ©berggeist007 / pixelio.de

More available books at **www.hansebooks.com**

EVOLUTION

OF THE

HUMAN RACE FROM APES

AND OF

APES FROM LOWER ANIMALS

A DOCTRINE UNSANCTIONED BY SCIENCE

BY

THOMAS WHARTON JONES, F.R.S. F.R.C.S.

MEMBER OF THE SOCIETY OF BIOLOGY OF PARIS, OF
THE IMPERIAL-ROYAL SOCIETY OF PHYSICIANS OF VIENNA, AND OF
THE ROYAL MEDICAL SOCIETY OF COPENHAGEN; PROFESSOR OF OPHTHALMIC
MEDICINE AND SURGERY IN UNIVERSITY COLLEGE, LONDON, AND OPHTHALMIC SURGEON
TO THE HOSPITAL; FORMERLY LECTURER ON PHYSIOLOGY AT THE CHARING
CROSS HOSPITAL; AND FULLERIAN PROFESSOR OF PHYSIOLOGY IN
THE ROYAL INSTITUTION OF GREAT BRITAIN; ETC.

LONDON
SMITH, ELDER, & CO., 15 WATERLOO PLACE
1876

TO

SIR ROBERT CHRISTISON, BART.

M.D. D.C.L. LL.D. F.R.S.E. F.R.C.P.

PROFESSOR OF MATERIA MEDICA IN THE UNIVERSITY OF EDINBURGH

PHYSICIAN IN ORDINARY TO H. M. THE QUEEN IN SCOTLAND

PRESIDENT OF THE BRITISH MEDICAL ASSOCIATION WHICH MET IN EDINBURGH IN 1875

PRESIDENT ELECT OF THE BRITISH ASSOCIATION FOR THE ADVANCEMENT OF

SCIENCE TO MEET IN GLASGOW IN 1876

ETC.

THE WORTHY SON OF AN HONOURED AND LEARNED FATHER

This Work is Dedicated

WITH SINCERE RESPECT AND IN TOKEN OF HIGH APPRECIATION OF

THE SCIENTIFIC AND PRACTICAL VALUE OF HIS LABOURS

THROUGHOUT A LONG AND DISTINGUISHED

PROFESSIONAL CAREER

ADVERTISEMENT.

THE FOLLOWING PAGES comprise two Lectures originally delivered to audiences of Ladies and Gentlemen in the Botanical Theatre of University College, London :—the first on the evening of October 29, 1874, and the second on the evening of October 12, 1875.

I now publish them, in the hope that they may serve as a means of counteracting efforts made throughout the country by half-informed persons to propagate the Doctrine of Evolution. Those efforts, it may be observed, are manifested under two different characters. Thus, while the more logical of the Propagandists referred to do not shrink from the conclusions which necessarily flow from the doctrine, a weak-minded class commit the absurdity of trying to reconcile the doctrine with Belief in a personal FIRST CAUSE.

Perhaps, the explanations herein offered will help to disabuse the minds of some, at least, of those who may have already been led unwittingly to accept the Doctrine, and will fortify with rebutting reasons those who, on the other hand, have instinctively recoiled from it.

In conclusion, let it be observed that, in contesting the scientific soundness of the Doctrine of Evolution, and arguing for the reasonableness of the common belief that all things were created by design and for a purpose, I necessarily refer to the Creator's wisdom and power, but I abstain from any theological discussion. Even if the conditions under which my Lectures were delivered had permitted, it would have been out of place to introduce Theological arguments, seeing that my aim was, as already intimated, simply to combat the claims set up for the scientific validity of the Doctrine of Evolution—not supererogatorily to defend Revelation.

It is commonly assumed by Evolutionists, and tacitly granted them, that there are only two fundamental alternatives between which a definite decision has to be made, viz.—Revelation or Evolution. But this I emphatically deny. Unbelief in Revelation and a personal Creator is no doubt the alternative of the admission of Evolution, but admission of Evolution is not the necessary alternative of Unbelief. Excluding all belief in Revelation and a personal

Creator, for argument's sake, I hold that the Doctrine of Evolution, unsanctioned as I believe I show it to be by Science, cannot be accepted as the alternative on any consideration, and must, therefore, be unconditionally and absolutely rejected.

PREFACE.

—◦◦◦—

'OUR present advanced knowledge in Natural
Science has not rendered the idea of Evolution a
bit more probable than it was in former times. And
it must be firmly denied that the conceit of " Natural
Selection by Survival of the Fittest " has, in any
degree, imparted to the theory more substantial body
than it had before, or raised it to the Scientific position which Darwin and his followers claim for it.'
Such are the words with which I conclude my second
lecture.

There are various processes in Nature to which
the idea of ' Natural Selection ' is logically enough
applicable, but it is amusing to observe how instances
of the kind have been illogically twisted, and adduced
as tests in proof of the soundness of the idea as
applicable to the Doctrine of Evolution, and even in
proof of the doctrine itself. Natural Selection, in the
sense in which it is applied to Evolution by Mr.
Darwin, let it be repeated, is a mere conceit. When,

therefore, it is adduced as the efficient cause of the supposed transmutations of lower into higher animal forms, one phantom is virtually conjured up as the cause of another.

The doctrine of Evolution in general, however, without any attempt to explain its cause, appears to be somewhat more defensible by itself. And for this reason, I think that Mr. Huxley, my friend and former pupil in Physiology at the Charing Cross Hospital, the ablest General perhaps whom the Brigades of the ' Evolutionary Army' can boast of, shows good strategy by manœuvring chiefly in defence of the position of Evolution pure and simple. Even Professor Haeckel, who pipes so lustily in praise of Darwinism that he might be appropriately viewed as Piper to the *Fairy-dance* described in the lines with which I conclude my first lecture, has for the burden of his tune : '*Phylogenesis* recapitulated in *Ontogenesis.*'

In his notice of Professor Haeckel's book on *Anthropogenesis* in the journal called the ' Academy' for January 2, 1875, Mr. Huxley adduces, as an example of what Geology teaches in favour of Evolution, *the actual historical process,* as he considers it, by which one species of animal now living, the horse, came into existence during the Tertiary epoch. ' The evidence,' says Mr. Huxley. ' based on the analogy of known developmental facts' (or, as Professor Haeckel would call them, Ontogenetic facts) that a 'three-toed Hipparion form, which lived in the Miocene epoch, gave rise by suppression of the phalanges of its rudimental toes and other slight modifications, to the apparently one-toed later tertiary horse is satisfactory to my mind —as satisfactory,' Mr. Huxley continues, as the evidence based on the analogy of known structural facts (that is, the known correlations of anatomical structure) which leads him to entertain no doubt that the said extinct Hipparion had a simple stomach and a certain kind of heart.

The two classes of facts here referred to, the *structural* or *anatomical* facts, and the *developmental* or *ontogenetic* facts, do not equally warrant by analogy the inferences which Mr. Huxley would respectively draw from them. To infer from an examination of two given skeletons of extinct animals, the anatomical characters of the soft parts of their bodies, which have disappeared from decay, is a simple matter of comparison and induction. It is, in fact, no more than saying that the fossil bones and teeth before you, present, the one set, the peculiarities of the bones and teeth of a horse or ass, for example; and the other set the peculiarities of those of a ruminating animal, such as a sheep, or ox; and that therefore, it is in the highest degree probable that the animal of which the one fossil skeleton is the remains belonged to the Equine family of the order *Pachydermata*, and that it had a simple stomach, with all the other general peculiarities of organisation which characterise that group ; whilst the animal of which the other fossil skeleton is the remains belonged to the order *Ruminantia*, and that it had a compound stomach, with all the other general peculiarities of organisation which characterise animals that chew the cud.

Whilst then it is a simple matter of comparison and induction to infer from an examination of the fossil skeleton of an extinct animal the anatomical structure of the other organs of its body which have disappeared from decay, it is the merest assumption to assert that a comparison of the observable mode of the embryonic development of an individual animal, with the various persistent forms throughout the zoological scale, supplies facts from which the doctrine of Evolution may be *legitimately* deduced.

In regard to the three-toed Hipparion form, we can recognise, with Mr. Huxley, in its bones and teeth, characters indicating a near alliance to the horse. We may also, on the other hand, recognise in the embryo of the horse indications—transitory indications of the phalanges of a would-be three-toe development, but to say, for that reason, that the

Hipparion was the ancestor from which the horse of the present day was evolved is an inconsequence; the two questions being altogether different.

The endeavour of Cuvier to construct from the study of fossil bones an anatomical and physiological history of the individual animal of which those bones are the sole remains, we thus see, was quite logical; but is wholly different in principle from the fallacious attempt to make the facts of Ontogenesis or individual embryonic development prove the validity of Phylogenesis, or Evolution of the line of all living forms by gradual increase and modification of structure throughout innumerable generations, in the course of millions of years, from a spontaneously produced shapeless mass of protoplasm, like a flake of the white of egg.

This leads me to the question of the influence of time. It is obvious that none of the metamorphoses which we can observe in the course of development of the young of animals going on before us, can be appealed to as having any bearing on the subject of Evolution other than that of Ontogenesis as the alleged recapitulation of Phylogenesis, so fully considered in my second lecture. Seeing that generations and generations innumerable are appealed to as constituting fundamental conditions for the transmutations of Evolution, the variations undergone by Mr.

Darwin's pigeons in the course of a few generations
are nothing to the purpose—nothing to the purpose
certainly, if the three or four thousand years comprised in Egyptian records cannot be admitted as a
sufficiently long time for the manifestation of evolutionary changes.

I have above alluded to applications which may
be logically enough made of the general idea of
Natural Selection in illustration of various processes,
and commented on the illogical reasoning which
would appeal to such instances as tests in proof of
the soundness of the idea of Natural Selection in its
application to the alleged Evolution of organised
beings. An example of such a mistake is presented
in the title of Schleicher's work : ' Darwinism tested
by the Science of Language.' The production of
language, however, to quote from Professor Dwight
Whitney, of Yale College, U.S., had nothing to do
as a cause with the development of man out of any
other or lower race. The only development in which
language was concerned is the historic development
of man's faculties. Language, therefore, in its beginnings, can be considered only as connected with
the history of man as man—not with any alleged
evolution from apes, as Mr. Darwin argues.

Mr. Huxley, assuming that the Doctrine of
Evolution cannot be gainsaid, observes : ' If the
evolution of all living forms, by gradual modification,

is an *historical fact* (?), why should the attempt to
reconstruct the details of that momentous history be
regarded as less philosophical or less laudable than
the attempt of a Niebuhr or a Mommsen to build up,
from ruined monuments, fragmentary inscriptions,
and obscure and often contradictory texts, a connected and intelligible history of Rome?' As to
attempts like those of Niebuhr and Mommsen here
referred to, they belong, it need scarcely be observed,
to a category altogether different from that to which
we must relegate the attempts of Evolutionists to
reconstruct the details of their 'momentous history'
of the origin of all living forms. Niebuhr and
Mommsen's building up of a connected and intelligible history of Rome has a parallel in Cuvier's
reconstruction of the anatomical and physiological
history of an extinct animal from the study of its
fossil remains ; but *no parallel* in any reconstruction
of an ideal history founded on assumption, such as
that which Mr. Huxley advocates. Here let it be
particularly understood, that by these words I of
course impugn the palpable facts of paleontology
which Evolutionists mix up with their history as
little as Niebuhr questioned the facts interwoven
with the myths in the old poetical legends of the
country.

Historical and ordinary anatomico-physiological

a

researches have much in common in respect to method— the object of both being to ascertain facts and to trace out their concatenations and correlations. In lately editing for the Camden Society the life and times of my ancestral kinsman, brave old Bishop Bedell, of Kilmore— 'Saint, Sage and Hero' as he has been happily characterised by Mr. Gladstone, I followed, in a small way, the example of Niebuhr and Mommsen, inasmuch as I searched or had searches made in all directions, for original documents —wills, state papers, letters, and the seals thereto attached, parish registers, diocesan registers, tombstones, &c. In doing so, the aim was to obtain *authentic* information from *genuine* contemporary documents, in regard to the true reading of which there could be no doubt, and thereby to correct various current misconceptions. This, it is obvious, was merely the prompting of common sense and duty— although I believe that habits of scientific method in physiological and pathological researches may have materially aided the exercise discrimination in my genealogical and historical inquiries.

T. WHARTON JONES.

35 GEORGE STREET, HANOVER SQUARE, LONDON :
January 1, 1876.

CONTENTS.

LECTURE I.

LECTURE II.

LECTURES.

———◆◆◆———

LECTURE I.

THE THEORY OF EVOLUTION IN GENERAL, AND MR. DARWIN'S DOCTRINE OF 'NATURAL SELECTION' IN PARTICULAR.

SUMMARY.

Like the Philosophy of Epicurus, the theory of Evolution excludes all idea of a personal FIRST CAUSE, p. 3. The doctrine of Evolution rests mainly on conceit and assumption, and is unsanctioned by science, p. 4. Alleged origin of man and animals by successive evolution throughout a long period of time from some simple living being which had itself been spontaneously generated by the assumed operation of natural physico-chemical laws, p. 4. Evolution a very ancient idea, and one which has often been reproduced in a more or less modified form, p. 4. Lamarck's scheme of Evolution published in France between seventy and eighty years ago, p. 5. Mr. Charles Darwin's objection to Lamarck's scheme as affording no explanation in respect to the mode in which the alleged process of evolution took place, p. 6. Evolution in its very first principles is a reproduction of the old Epicurean doctrine as recited by Lucretius in his poem, 'De Rerum Natura,' p. 6. True Evolutionists agree with Lucretius and Descartes in denying design in Nature, p. 7. Mr. Charles Darwin's doctrine of 'Natural Selection' in explanation of the mode in which the alleged evolution of higher from lower grades of organisation has taken place, p. 7. Such a doctrine is fundamentally no more than what Epicurus taught in Athens, and Lucretius after him sang in

B

Rome, and is a mere truism so far as there is any meaning in it, p. 8. The doctrine of Evolution by 'Natural Selection' is now so associated with Mr. Darwin's name, that it has come to be designated 'DARWINISM.' Nay, the doctrine in general is now confounded under that name, p. 9. According to the doctrine of Evolution, Mankind are merely the sons and daughters of Apes ; which latter were themselves descendants, through the lowest mammals, of frogs and fishes ; whilst those fishes came from an ascidian mollusk, which itself descended, through worms and protozoa, from a spontaneously generated cytode, p. 10. The time occupied in this assumed series of evolutions has been inconceivably long ; but to speak of it as indicating the antiquity of man is to confound in the calculation the altogether different question of the antiquity of Man as Man, p. 11. Mr. Darwin thinks that the fundamentally similar construction of the human arm and hand, for example, the fore legs and feet of quadrupeds, wings of bats, flappers of seals, paddles of whales, and wings of birds, inexplicable on any other supposition than common descent, p. 11. But such similarity of plan in construction does not logically prove that animals must have been evolved in the manner alleged, p. 12. That every thing was created by design and for a purpose is perfectly explanatory to common sense, and in no way contradicted by Science, p. 13. Rudimentary organs alleged to be ' instruments without use,' but they are certainly in no case without a use of some kind in the economy in which they occur, p. 14. The limbs and lungs of the frog come into use when they come into existence, while the tail and gills of the tadpole fall out of use in proportion as they shrink and disappear. This takes place according to a law of the economy of the frog, not by use or disuse, p. 16. According to the Evolution doctrine, it depended on the chance of circumstances whether the pretended transmutations of structure took place in this or that manner, p. 21. The transitional resemblances observable in the form and structure of animals from worms up to man indicate *not evolution* but *unity of design* and *designer*, p. 22. To Evolutionists, however, transitional resemblances are inexplicable except on the assumption of lineal descent, p. 22. Homologies of structure have been long ago traced out in more or less detail, but their demonstration does not in any way really help the doctrine of Evolution, p. 24. As to the alleged evolution of man from apes and the affinity between him and them, it is to be observed that though the tail-less apes have a counterfeit presentment of the human form, they are unmistakably brutes, and do not in respect to mental powers rise above other brutes lower in the zoological scale, much less make the most distant approach to man, p. 25. Mind can never have been evolved from a mind not in existence by means of any absurdly alleged exertion of the nonentity, p. 27. Unfounded nature of the assertion, that the peculiarly endowed brain-cells of man could ever have been evolved from differently endowed brain-cells of lower animals, p. 28. Any increased mental activity of an ape could never have led to the evolution of the human mind, p. 29. According to the theory of Evolution, function determined structure, but this is an absurdity. The idea that function preceded mechanism is virtually admitting design, but shifting it from the *Creator* to the *Creature*, p. 30. From the first, Man was endowed with moral and intellectual faculties, and a capacity for development and improvement, p. 30. Advance in culture means merely that the inherent capabilities of man's nature have been *called forth by education*, not *acquired by evolution*, p. 31. No valid grounds for the thesis that man ever existed under any other *embodiment* than that of MAN, p. 32.

LADIES and GENTLEMEN,—Amongst those who encountered St. Paul at Athens and sneeringly asked 'What will this Babbler say?' were certain philosophers of the Epicurean sect.

Like the philosophy of these same Epicureans, the Theory of Evolution excludes all idea of a personal Creator—all idea of design—and all idea of a moral order in the world.

Involving, as it thus does, the tremendous issue of sapping the foundations of religion, such a theory, to deserve being entertained, ought, at least, to be constructed on facts the most exact and pertinent and arguments the most logical and conclusive. There ought not to be adduced in support of it any conceits or assumptions under the guise of inductions alleged to be legitimately drawn from scientific investigation. But I do not hesitate to maintain that, instead of the conditions of true science, the doctrine of Evolution rests mainly on conceit and assumption. Its inductions have far outrun facts, and the conclusions deduced from those inductions are recklessly sweeping.

While no pious plea, or any other consideration, ought to be permitted to influence the conclusions which may be justly and rigorously deducible from scientifically established truths, we must resist all speculations, how much soever vaunted, which it may be sought to impose upon us in the name of science—*though without the sanction of science*—and by the assumed authority of peculiarly constituted minds.

The remarks I am about to make will show how far we may be justified in rejecting such speculations as those comprised in the Theory of Evolution.

Man and animals were not originally created in their present form, but have been evolved, in the course of successive descent, by gradual growth, development and transmutation of structure, throughout a long period of time, from some simple living being which *itself* had been spontaneously generated according to natural physico-chemical laws :—Such is the general expression of a very ancient idea, and one which has often been reproduced in a more or less modified form. Not to

go further back than the end of the last century and
beginning of the present—the doctrine was advo-
cated by Dr. Erasmus Darwin in England, and
Lamarck in France. The latter it was who—in his
'Philosophie Zoologique'—drew out a scheme of
evolution in a systematic form and in all its conse-
quences.

Lamarck's conclusion was, that Nature, in pro-
ducing successively all the species of animals, and in
commencing with the most simple, to complete her
work with the most perfect, has gradually added to
the complexity of their organisation. Under the
influence of the circumstances into which these
animals fell, in their distribution all over the habitable
globe, each species has acquired the habits belonging
to it, and the modifications in its parts, which observation shows it to possess. Everything, therefore,
argues Lamarck, concurs in *supporting* his assertion,
that it is not the form of the body or of its parts
which gives rise to the habits and mode of life of
animals ; but on the contrary, that it is the habits,
mode of life, and other influences, which have, in the
course of time—ages and ages—determined the form
of the body and parts of animals. With new forms
new faculties have been acquired, and Nature has
gradually come to fashion animals such as we now
see them.

Mr. Charles Darwin, the grandson of Dr. Erasmus Darwin before referred to, objects to Lamarck's
scheme of evolution, which I have just recited, as

affording no explanation of the mode in which the alleged progressive transmutation of organic bodies from the lowest to the highest grades has taken place.

Before proceeding to consider Mr. Darwin's own edition of Evolution, let me stop to declare more particularly what the doctrine, under whatever aspect or form it may be viewed, logically implies in regard to Nature and the non-existence of God. The doctrine of Evolution excludes, as I have before said, all idea of a personal Creator. According to it, all that we have been accustomed to regard as evidences of design, and as the work of an Almighty, Wise, and Beneficent God, are the result merely of the spontaneous and necessary operation of the inherent properties of matter.

Evolution, we thus see, is in its very first principles virtually a reproduction—improved, indeed, by a more advanced science—of the old Epicurean doctrine recited by Lucretius in his poem 'De Rerum Natura.' 'It was not by design,' said he, 'that atoms framed the world; but after many fruitless collisions they chanced to fall into such motions as produced the world and all that is in it. At first monsters of all kinds were formed which could not grow up nor continue their kind. They all, therefore, perished off.' ' The members and organs of the body,' further said Lucretius, 'were not formed by design, but, having been formed, they came to be applied to the uses for which they were found adapted.'

Matter, according to the advocates of the doctrine of Evolution, is eternal ; and natural physico-chemical forces are, as I have stated their contention to be, the causes of the phenomena hitherto ascribed to a Supernatural creative power. There is, they say, with Lucretius and Descartes, no final cause or design in Nature ; nor any moral order in the world. On the contrary, everywhere there is war of all against all—everywhere a struggle for existence.

To come now to Mr. Charles Darwin's own views as to the mode in which the alleged progressive transmutation of organic bodies from the lowest to the highest grades has taken place. The efficient cause in operation Mr. Darwin considers to have been what he calls ' Natural Selection '—a process described by him as follows :—' Amid the struggle for existence which has been always going on among living beings, variations of bodily conformation and structure, if in any degree profitable to an individual of any species, will tend to the preservation of that individual, and will generally be inherited by its offspring.'

Thus is ' Survival of the Fittest ' the condition for ' Natural Selection.'

A fundamental point in the doctrine of Evolution, as I have before intimated, is that it has been only very gradually that the alleged growth, development, and transmutations of organic structure and form have taken place from generation to generation, and that, therefore, it has been only in the course of long ages of time that the animal kingdom has attained its present state of organisation, by evolution from a hypothetical spontaneously generated living being, consisting of no more than a structureless mass of protoplasm, like a flake of the white of an egg.

This point in the doctrine of Evolution, it is obvious, renders impossible anything like a just comparison of its import with the facts which our minds can, in consequence of our very brief tenure of existence and limited experience, properly grasp. We may, however, concede to Evolutionists all the millions of years which they postulate, and yet decline to receive their doctrine.

But to return to Mr. Darwin's 'Natural Selection.' The preservation of favourable and the decay or lapse of injurious variations, constituting 'Natural Selection' by the 'Survival of the Fittest' in the 'Struggle for Existence' seems no more than what Epicurus taught in Athens, and Lucretius, after him, sang in Rome, before the commencement of our era; and is a mere truism so far as there is any meaning in it. The more healthy and well developed individuals are, they have, no doubt, a better chance of surviving and continuing their kind than the less well developed; though it is difficult to perceive that they would be thereby elevated above the condition and degree of organisation proper to their race. Could survival of the fittest lead to more than an improvement of the breed, and eventually, perhaps, to a variation of the species ? But are not variations prone to disappear, and a recurrence to the original type ensue ? And what, it may be asked, constitute favourable and what injurious variations ? The acquisition, to answer the question broadly, by one kind of animal of a structure or organ proper to another kind, though of more perfect development, would not be a favourable variation. For, as Lucretius himself admitted, 'thence would rise vast monsters.'

The idea of 'Natural Selection by Survival of the Fittest' is thus—as well as the first principles of Evolution—clearly implied in the Epicurean Philosophy. 'Of the multifarious beings formed by the fortuitous concourse of atoms,' said Lucretius, '*the fittest alone survived.*' 'Many races of regularly organised creatures'—to quote again from Monro's excellent translation of Lucretius—'must have died off because they wanted some natural power by which to protect themselves.' * * * 'These fell a prey to others, and disappeared unable to endure *the struggle for existence.*'

The 'Natural Selection' theory, though thus foreshadowed about 2000 years ago by Epicurus and Lucretius, is now so associated with the name of Mr. Charles Darwin—as is even the doctrine of Evolution in general—that Evolutionists look up to

him as the prophet of their faith, and boldly assert
that he has done for Biology what Newton did for
Astronomy! In regard, however, to this comparison
of Natural Selection with Universal Gravitation, it
may be remarked that we look in vain for any real
basis to the conceit of Natural Selection, like the
accurate measurements of Picard and the profound
calculations of Newton, from which the grand law of
UNIVERSAL GRAVITATION was deduced.

According to the principles of Evolution, which
I have thus briefly sketched, MANKIND are merely
the Sons and Daughters of APES ; and if we trace
our ancestors still further back through the lowest
mammals to frogs, and thence through fishes to
invertebrates, we shall probably find among the
latter, our two great-grandparents—in what degree,
shall I say—the million-millionth ? rolled into one,
in the shape of an hermaphrodite ascidian mollusk,
molluscoid, or worm—whichever you choose to call
it ;—whilst, passing back from this ascidian progenitor through protozoa, we come in the last
instance to our alleged FIRST PARENT, a *spontaneously
generated cytode !*
Such is, in bare outline, the genealogy of Man,
which has been propounded with the claim of being
a special deduction from the Evolution theory.
Evolutionists, as I have said, claim a very high
antiquity for the human race. It appears, however,
difficult to understand in what way the argument

can be supported by the concession of this postulate; for as Man's history could have begun only at the time of his first appearance on the earth, it is immaterial whether the period of his existence thereon be reckoned by thousands or by millions of years. The scientific question of his origin would remain the same. But Evolutionists, in their claim of a very high antiquity for the human race, mix up this question of the antiquity of Man as Man with the other and altogether different question, viz. that of the whole time occupied in the assumed series of evolutions, from a protogen to an ascidian mollusk, and from an ascidian mollusk up to a human body.

Into the composition of animal bodies certain structures and organs enter as fundamental materials; but in accordance with the particular nature of the different kinds of animals, they are variously modified, as regards both the degree and character of their development.

Of the homologous, or similar fundamental materials, for example, which compose our arms and hands, the fore legs and feet of quadrupeds are formed, the wings of bats, the flappers of seals, the paddles of whales, the pectoral fins of fishes, &c. By a somewhat different modification of the same materials also, the wings of birds are constructed.

According to Mr. Darwin this fundamentally similar composition of the human arm and hand, the bat's wing, the whale's paddle, &c., is utterly inexplicable on any other view than descent from a common progenitor, together with subsequent adaptation to diversified conditions. 'Thus,' continues Mr. Darwin, 'we can understand how it has come to pass that man and all other vertebrate animals have been constructed on the same general model, why they pass through the same early stages of development, and why they retain certain rudiments in common.' 'Consequently,' he adds, 'we ought frankly to admit their community of descent;—to take any other view is to admit that our own structure, and that of all the animals around us, are a mere snare laid to entrap our judgment,'—'*are a mere snare*'—let me repeat the words,—'*laid to entrap our judgment!*'

In like manner, Mr. Darwin might have said that all our intuitive ideas respecting the existence of God must be, and are, mere snares laid to entrap our judgment. But by whom or by what laid? let me ask. By the operation of natural physico-chemical laws?

To Evolutionists who adopt such a dictatorial mode of arguing as this, no other answer ought to be given than that of Job to Zophar the Naamathite :—' No doubt but ye are the people, and wisdom shall die with you.'

The similarity of plan in the construction of their organs does not logically prove that animals have been evolved from each other—the higher from the lower—as Evolutionists maintain. Affinities

and homologies of structure do not entrap my judgment at least, but are to me manifestations of the
unity of the Creator's plan.

Our knowledge of the existence of the material
world, let me add, is not more certain than our
knowledge of the existence of God the Creator.
While our sensations lead the mind by intuition to
believe in the existence of matter ; our mental perceptions as surely lead us by intuition to believe in
the existence of God.

But to return :—In the prevailing order of this
world, everything we see works together in conformity. It is, therefore, perfectly explanatory to
common sense, and in no way contradicted by
science, that everything was planned and created by
the same Divine mind and power. Admitting fully
that this is no scientific explanation, I maintain that
it is equally no scientific explanation gratuitously to
assume the dogma, that peculiarities of conformation
in animals depend on inheritance from a progenitor
which had, by chance, the part of the body possessing the peculiarity of conformation congenitally
modified in structure so as to be different from that
which was proper to its earlier progenitors.

Rudimentary organs, according to the doctrine
of Evolution, are ' instruments without use '—' parts
without anything to do '—and which present themselves merely because they have been inherited
from a progenitor in which they existed fully

developed and capable of action! 'In order,' says
Mr. Darwin, 'to understand the existence of rudimentary organs, we have only to suppose'—let me
repeat the words—'*we have only to suppose* that a
former progenitor possessed the parts in question in
a *perfect state*, and that under changed habits of life
they became greatly reduced, either from simple
disuse or through the natural selection of those
individuals which were *least incumbered* with a
superfluous part, aided by the other means previously
indicated.'

Many such rudimentary structures Mr. Darwin
affirms to be neither beneficial nor injurious to the
animal in which they exist; and the probability is,
according to him, that all organised beings, including
Man, possess many modifications of structure of no
service either now or formerly!

That an organ even of the least developed character found in any one animal is of no use in the
economy of that animal, but has been merely
reproduced in what is called a rudimentary form,
because the organ it represents had existed in a
more fully developed state in some alleged remote
ancestor, I do not hesitate to pronounce *a mere gra-
tuitous assertion*, the hasty assumption characteristic
of *amateur anatomy and physiology*. How rudimentary soever an organ may be and how unim-
portant soever, comparatively speaking, and though
it does not perform the function proper to its
homologue as more fully developed in other

animals, it is still in my opinion not without a use
in the economy in which it occurs. The small
wings of the wing-powerless birds, for example, are
not useless, though they do not enable the animals
to fly. The wings of the ostrich aid it in running;
the wings of the penguin serve as paddles in swimming.

What Mr. Darwin, following Lamarck, says of
the use and disuse of organs—*use* as the cause
of increased development, *disuse* as the cause of
deterioration of structure—appears to me to be the
result, in a great measure, of *ill-observed* and *ill-
explained facts*, and therefore no argument in support
of his doctrine, as he *illogically* imagines.

That animals have had their form and structure
and, consequently, their functional endowments
transmuted by Natural Selection or any other process
of Evolution in a long course of ages, and that they
have thereby become adapted for habits and a mode
of life different from those for which their ancestors
were originally created, are propositions not supported
by any legitimate induction from real facts.

Very exaggerated and illogical notions pass
current as to the effects of use and disuse on organs.
Enlargement and increased physical strength are
considered the proper effects of use; diminution of
size and impaired physical strength the proper
effects of disuse; whatever be the nature of the
organ in question, though this is really applicable

chiefly to bones and muscles. But enlargement with increased physical strength, and diminution of size with impaired strength, are quite different things from *transmutation*.

Organs are used because they are fit for use ; organs are not used because they are unfit for use. We can, for example, observe how the limbs and lungs of the frog come into use, and how the tail and gills of the tadpole fall into disuse ; and how the animal from being virtually a fish comes thereby to be an air-breathing Batrachian.

The limbs of the young frog thus metamorphosed come into use because they have come into existence under all the conditions fitting them and exciting them to perform their appropriate actions. The animal in the tadpole state has no need of the limbs before they are formed ; and when they have been formed the animal must necessarily use them, because the conditions for their action have come into existence along with themselves.

Let us now see how an organ falls out of use and disappears. This process we can observe in the tail of the tadpole. That organ does not disappear because it ceases to be used, but ceases to be used in proportion as it shrinks and disappears ; while the fore legs of the future frog become, like its already developed hind legs, free and fit for action.

These changes constitute a stage in the natural process of development to the perfect frog-state, when the respiration by gills is at the same time superseded by the completed development of the lungs. The gills disappear not because they cease to be used, but they cease to be used because they shrink and disappear by a law of the economy of the frog; whilst by the same law lungs are developed and the animal breathes air. Lungs thus come into use *because* they have come into existence, *not because* the animal first accustoms itself to live on land and breathe air.

It is when the fore legs become free and the animal thereby fitted for progression on land, that the tail shrinks and disappears, and the gills for respiration in water are superseded by lungs for breathing air. It is not until the fore legs have become developed that the young frog is able to make its way out of the water to breathe air. If by accident the animal, now in the frog phase of development, cannot get a footing on dry land, it is liable to be drowned just as surely as the tadpole was to die from asphyxia, like a fish, if taken out of the water. No sooner does the young frog reach dry land than it begins to jump, its strong hind legs being already fitted for the purpose without any previous practice, except in swimming before it left the water.

We thus see that in the process of metamorphosis the young frog is structurally prepared for its new conditions of life. The tadpole is not first placed in the new conditions and then has its structure changed under the influence of those conditions in order to be adapted to them. All parts of the structure of the frog being in conformity, one change in the

c

organisation of the tadpole is necessarily accompanied by others—originating certainly *not by chance, but by design.*

The large and heavy birds of oceanic islands, says Mr. Darwin, have wings so rudimentary that they are incapable of flight; and this he explains by supposing that the progenitors of those birds not having been exposed to the attacks of beasts of prey, were not necessitated to fly, so that their wings came at last by this disuse in the course of generations so reduced in size as to be unfit for flight. On this I would observe, that though not necessitated to fly to escape the attacks of beasts of prey, the birds, supposing they had ever been capable of flight, would have been necessitated to exercise that faculty in looking out for food. The power of using their wings would not, therefore, be likely to have come to be lost.

But, let me ask, why suppose those wing-powerless birds to have been descendants of birds with well developed wings? Would it not be easier, according to the principles of the Evolution doctrine, to suppose —if we must suppose anything—that the first evolved birds possessed mere rudiments of wings, and that the birds of oceanic islands were descended from them without increased development of wing.

That organs have become rudimentary by degeneration from parent to offspring in the manner Mr. Darwin here assumes, appears to me quite a fancy. There is no proof whatever—no reason of any kind to suppose that an organ which we now find rudimentary in any given animal was ever at any time in the progenitors of that animal other than rudimentary. An animal in possession of wings fitting it to fly will fly. The young bird when fledged will essay its wings like as the young frog I have mentioned does its legs in jumping on emerging from the water, being impelled thereto by the internal feeling which the possession of wings imparts. ' The new fledged offspring' requires not to be ' tempted to the skies' by the old bird. Ducklings hatched by a common hen do not imitate their foster-mother, but in obedience to their own instincts, which depend on their whole organisation, rush to the water and swim.

If, it may here be remarked, one organ could, by disuse and natural selection, become rudimentary, as Mr. Darwin assumes, all the organs of animals one after another might by chance of circumstances and in the course of time become more and more rudimentary, so that the descendants of organisms of high development might, thereby, become eventually reduced to the protozoon-state, whence they are alleged to have been originally evolved !

Dexterity in the exercise of the function of an organ has been often attributed exclusively to increased power of the organ itself from practice, whereas animals, as we have seen, make use of their organs as well at first as afterwards;—and as regards man, his dexterity for a purpose really depends, in the greatest degree, on the individual improvement, by education and experience, of the logical cunning of his mind in the exercise of his organs.

In exemplification of such erroneous views as to the acquirement of increased power by organs from practice, I may instance the common notion that the eyesight of civilised men is inferior in acuteness to that of savages. Among civilised men, numbers are endowed with eyesight as acute as can be for any distance,—far or near. The training of the eyesight for any particular purpose is a different question, and means the education of the mind to exercise the sight and to interpret the particular visual perceptions. Those savages who have first-rate emmetropic sight do not see better than civilised men who also possess emmetropic sight, that is, who possess eyes with an optical conformation and adjusting mechanism as perfect as can be for vision at any distance ; but being more familiar with his hunting ground and the objects there met with, the savage can recognise the objects—*where they are and what they are*, and *the traces of them*, better than the untrained civilised man, at great distances. On the other hand, it will be found that savages, though they may see near and minute objects well enough, cannot distinguish them and appreciate their various details so well as civilised men who have been much practised in the nicer examinations and manipulations

common in civilised life.   But to return from this
digression.

In the natural history of an organism we recognise something more than the manifestation of
physical and vital forces—physical forces as the
attribute of the machinery—and vital forces as
the mainspring of its action.   We recognise in the
aggregate plan a Divine Idea, and in the fulfilment
of the purpose an Almighty Hand.   From the commencement of its development the body of an animal
is in a continuous process of change, and yet it
remains unchanged in plan.   The transformations
merely run in a circle, so that there is no progressive
evolution ; but the different races, so long as they
exist, continue to retain each its own characters.

Our experience does not permit us to admit more ;
nor can we logically infer more from any existing data.

In the production of the elaborate mechanism of
organised and living beings, manifesting to ordinary
comprehension design for the fulfilment of a purpose,
Evolutionists, as before stated, exclude the idea of a
Creator and attribute it all to chance, which is really
the meaning of *Natural Selection.*   Or, if organised
and living beings were not exactly and directly
evolved by chance, they were produced in some
such way as this :—At an incalculably remote period
a living body was spontaneously generated, under
the operation of mechanical or physico-chemical
laws,—a body of extreme simplicity of composition

and structure, indeed, but so marvellously endowed
that it comprised within itself the potentiality of
passing through, *in its descendants*, a series of metamorphoses, in the course of long generations, of this
or that kind according to the chance of circumstances.
The eventual, though unpreordained, result of which
has been the vegetable and animal kingdoms such as
we now find them.

Even on a superficial survey we may recognise
transitional resemblances in the form and structure of
animals from worms up to Man; while a deeper insight
discloses in the plan of their organisation such a unity
of design, that the whole animal kingdom appears
like one chain of beings successively more and more
developed. The chain, speaking of animals as we
find them living on the earth at the present day, does
not, indeed, appear to be an unbroken one throughout. Where, however, interruptions occur, connecting links may, in many cases, be traced in the fossil
remains of animal forms which are now extinct.

The general gradation in form and structure, and
the unity of plan in question appear to Evolutionists
inexplicable, except on the assumption that the higher
are lineal descendants through a long series of generations of the lower animal forms. This alleged inex-
plicability, however, appears to me to exist not in the
thing itself, but in the peculiarly constituted mind
which entertains the idea. No scientific reason can
be adduced in support of the doctrine of a progres-

sive transmutation of forms by descent from the lowest
to the highest grades. It is not a legitimate induction
from the facts of the case, nor a deduction from any
established principles. Seeing that the conditions of
life are fundamentally similar for all living beings, it
would be surprising, indeed, if there was not a unity
of plan recognisable in their organisation. We can
conceive of nothing more perfect than what we find,
and anything conceivably different would have been
less perfect. Every living thing—extinct or existing
—in short, has held or holds a place important in the
plan of the Almighty, Wise, and Beneficent Creator.

The transitions of form—the homologies and
analogies of structure—observable in the organic
kingdoms cannot be admitted, then, as indicative of
any such thing as a progressive development, whereby
animals have by chance or spontaneously, or by
Natural Selection—whichever you choose to call it—
been transmuted from a lower to a higher grade of
organisation. Among the astonishingly numerous
and varied forms of animal life, a certain resemblance
we have seen may, indeed, be traced indicative, as
before observed, of a unity in the conception and
design of the whole,—the same creative power being
manifested in the worm which is so magnificently
displayed in the human body. But though all organic
forms thus resemble each other, no two kinds are
exactly alike. There is a line of demarcation by
which each kind or form is, and has been, circum-
scribed.

An exposition of the affinities, homologies, and adaptations of structure throughout the animal kingdom, is a great aim of Philosophical Anatomy. And the subject has been long ago pretty successfully worked out—even to the tracing of transitions between invertebrates and vertebrates. But how well made out soever the affinities, homologies, and adaptations of structure may be, their demonstration does not help to confirm the doctrine of Evolution. The question is one beyond the pale of science, and any addition of facts will still fail to fill up the proof.

If thus we cannot discover sufficient evidence that the lower animals were evolved from each other, we might well abstain from any further discussion of the subject of Evolution. But, in pursuance of my design, I invite you to study with me the Nature of MAN ; and in doing so we shall examine the facts and arguments which have been adduced in support of the alleged evolution of the human race from Apes.

Whilst some minds are quick in detecting differences, others more readily perceive resemblances. It is a great aim with naturalists to detect specific differences so as to add new species to their catalogues; whereas comparative anatomists seek rather to discover resemblances in structure that they may thereby enlarge the domain of their doctrine of homologies. But as in their eagerness for discovery some persons are prone to draw a distinction without

a difference, so there are others who will recognise
an identity without a resemblance. A tendency to
this latter turn of mind seems to manifest itself when
it is contended by Evolutionists that the higher apes
are more closely allied to man than they are to the
lower members of their own order. That the orang,
chimpanzee, gibbon, and gorilla have a counterfeit
presentment of the human form is true, but still in
every single visible character they are unmistakably
brutes ; whilst as regards internal quality, they exhibit
not the most distant approach to man. Nay, they
cannot even be said to excel many mammals of a
rank below that of the order to which they are
referred in zoological classifications.

The numerous points of structure in the human
body which can be traced as homologous with points
of structure in the bodies of the lower animals, have
all been long recognised as evidences that man and
the lower animals have been constructed according to
one common plan ; but, let me repeat, they afford no
proof that man has descended from apes.

Considering, as before observed, the general
conditions of animal life in this world, and considering the perfection of every creature after its kind and
in its own sphere of life, we cannot conceive a reason
why organisation should have been different from
what we find it to be in man, or in any other animal,
nor any reason to infer from the similarity of organisation which prevails among animals that the higher
must have been evolved from the lower—that man

has descended from apes, apes from lower mammalian forms, or that the lowest mammals have been evolved through still lower vertebrates from invertebrates.

In regard to the Mental Powers of Man and the lower animals, Evolutionists assume that the difference, great as it is, is in degree only, not fundamental ; and in respect to connecting words with definite ideas, Mr. Darwin says, that ' it does not appear altogether incredible '—mark the argument, ' *not incredible !* '— ' that some *unusually wise ape-like animal* should have *thought* of imitating the growl of a beast of prey, so as to indicate to his fellow apes the nature of the expected danger.' ' This,' he continues, ' would have been a first step in the formation of a language.' On this I would remark, that to give point to such a supposition, we must, at the same time, assume it as ' not incredible ' that the fellow apes, to whom the *growl* was addressed, were able to understand the meaning of it, and, therefore, already as wise as the *growler*.

Mr. Darwin admits that ' the mental powers of some early progenitor of man must have been more highly developed than those of any existing ape, before even this most imperfect form of speech could have come into use !' No doubt of it, I say, though we must look upon any such early progenitors of man as imaginary beings, like the *Yahoos* of Gulliver's travels.

In regard to the curious argument of a relation

between the continued use of language and the development of the brain, it may be asked : How could the continued use of language, admitting such for argument's sake, have reacted on a *mind not in existence*, by enabling and encouraging the *non-entity* to carry on long trains of thought ? But in reality the use of language at all presupposes, as before observed, already existing intelligence. For as James Harris, in his ' Hermes' remarks :—'If, like lower animals, men had been by nature irrational, they could not have recognised the proper subjects of discourse.'

The formation and development of language—call it evolution if you will—is a question quite apart from that of the alleged evolution of the human race from apes. The question belongs only to the history of Man as Man.

In accordance with the Evolution doctrine, that the mental faculties of man, high as they are, have only been gradually developed from the more lowly faculties of his progenitors, the brutes, Mr. Darwin says :— ' In each member of the vertebrate series the nerve-cells of the brain are the direct offshoots of those possessed by the common progenitor of the whole group. It thus becomes *intelligible*,' he says,—let me repeat the word ' *intelligible*,'—' that the brain and mental faculties should be capable, under similar conditions, of nearly the same course of development, and consequently of performing nearly the same functions.'

The peculiar brain-cells of man, it is to be

observed, in reply to this statement, through which his high intelligence is manifested, have no known prototype or rudimentary representatives in the lower animals ; and as *nothing* can come of *nothing*, so it is impossible to admit that the human brain could ever have been evolved in the course of any number of generations from that of an ape or other brute, in which there is no manifestation of anything like human mind even in the most rudimentary degree.

In regard to cells generally, it is important to remember a fundamental principle in their biology— *a fundamental principle overlooked by Evolutionists*, —viz., that they are of various kinds, each kind possessing its own peculiar vital endowments, and its own mode of further development. One kind of cells cannot give origin to another kind of different endowments. There is no such thing as ' differentiation,' in the sense in which Evolutionists employ the term. As, therefore, the cells from which the optic nerve, for example, is developed cannot be developed into the auditory nerve, nor the cells from which the auditory nerve is developed, into the optic nerve ;— moreover, as the optic nerve cannot perform the function of the auditory nerve, nor this the function of that ; so, as none of the brain-cells of an ape can perform functions like those performed by the peculiar cells of the human brain, through which man's high intelligence is specially manifested, there is no reason to believe that any of the brain-cells such as

are possessed by apes could have come by repetition
of generations to acquire the endowments fitting them
for the performance of the highly specialised functions proper to those cells of the human brain in
question.

How, therefore, can it be maintained that by
increased mental activity reason could have been
acquired by apes through Natural Selection aided by
inherited habit ? There might, indeed, be increased
mental activity in the ape, but that would be, nevertheless, nothing more than *ape mental activity*, not
man's even in the slightest degree ; and what does
not exist cannot be inherited. The more an ape's
mental activity is increased, the more intensely *apish*,
indeed, it must prove.

In accordance with the teaching of the theory of
Evolution that function determines structure, Mr.
Darwin argues that as the mental faculties were
gradually developed, the brain would become larger.
But surely this is putting the cart before the horse,
for the truth is that the mental faculties are manifested only in proportion to the existing size and
quality of the brain. Function cannot precede
mechanism. The rails had to be laid down, and the
steam-engine put into working order, before we could
be conveyed from London to Brighton in an hour
and a half. The telegraph cable had to be submerged
in the Atlantic before messages could be flashed
between London and New York.

This view as to function determining structure
is, we have seen, associated with the direct denial of
design in the Creation; though, oddly enough,
design is implied in the doctrine of Evolution, only
it is shifted from the Creator to the living beings
themselves—from the Creator to the creature.
Thus, according to Dr. Erasmus Darwin, the acquisition of new parts in the course of its development
was the result of the animal's own exertions to
obtain what it longed for ; or, according to Lamarck,
new organs having become necessary in consequence
of new wants, efforts were made by the animal to
acquire the desiderated organ, and so it came into
existence! Lamarck even represents the hypothetical immediate ape-progenitors of man as planning
and executing the various steps requisite for advanc-
ing themselves to the dignity of manhood !

Against such views, which were upheld by
Geoffroy St.-Hilaire, in the memorable discussion
at the French Academy of Sciences, in 1830, the
Baron Cuvier raised his voice and maintained, what
our common sense acknowledges, that living beings
were created by design and for a purpose.

Man was created with his faculties—intellectual
and moral—capable and ready in the individual to
be elicited under the proper conditions. The least
civilised man in early times could not have been
lower than some savages of the present day. Now,
the lowest savage individually, it has been found,

has a capacity of being raised by education, continued from childhood, to a degree of intelligence
not manifested by his kinsmen in their savage state
and to which, by no training, under any circumstances,
can brutes be brought to make the most distant
approach.

Advance in culture means merely that the inherent capabilities of man's nature have been *called
forth by education, not acquired by evolution.* It has
not been by any fundamentally improved development of his corporeal frame or mental capacity in
the course of generations that man has advanced to
his present stage of civilisation and knowledge, but
by the preservation, communication, and transmission
of experience, acquired in all the various ways of
life in successive generations. This power to pre-
serve, communicate and transmit the knowledge
acquired by experience is a grand and characteristic
attribute of man, the wisdom and experience of the
individual being thus not lost to society by his death.

In the earliest times known to history men
existed with mental endowments as great as those
for which the most eminent men of modern times
have been distinguished, but they had not the
advantage of the same amount of accumulated
knowledge on record from which to start. The
mental capacity of man now is not greater than it was
some hundred years ago, and yet his achievements in
science, discovery, and invention within that time have
been unparalleled in the history of any former period.

As just said, man as we know him in his earliest records, industries, and arts—Egyptian, Assyrian, Indian, and Jewish—was as' well endowed formerly as he is now. His mental capacity was as great, and his corporeal frame as perfect. No doubt, contemporary with those civilised communities, there existed then, as now, tribes like the Australians, Papuans, Andamanese, Esquimaux, Fuegeans, &c., here and there, lost perhaps in barbarism, but nevertheless possessing all the capacity for being educated. Conversely, the remains of man and his industry which have been discovered under geological conditions undoubtedly denoting very remote antiquity, are, perhaps, not to be viewed as indications that prehistoric man was everywhere in the apparently low, savage, isolated state in which the people evidently were, whose remains have been discovered, for, possibly, contemporary with them there may have been civilised communities *elsewhere* on the earth.

The question of the *origin of Man*, it must be concluded, is one entirely beyond the pale of Natural Science. But when, where, and howsoever his first advent on the earth took place, this much is certain, . that there are no valid grounds in support of the thesis that he ever existed under any other presentment or embodiment than that of *Man*.

And now in conclusion : a great aim of Philosophical Anatomy, we have seen, is to discover and elucidate the Affinities, Homologies, Analogies, and Adaptations of Structure throughout the Animal and Vegetable Kingdoms ; but I have observed that, how well made out soever the subject may be, the demonstration of Affinities, Homologies, Analogies, and Adaptations of Structure does not add to the confirmation of the Doctrine of Evolution ; nor does the Doctrine of Evolution necessarily follow, in any respect, the demonstrations of Philosophical Anatomy. The two, however, having become associated by recent teachings in the minds of students, younger naturalists, who are impressed more by the surface than the depths of things, see in every homology of structure a manifestation of Evolution by Natural Selection, by Survival of the Fittest, in the Struggle for Existence.

As the Schoolmen of the dark ages lost themselves in trying to make of the Philosophy of Aristotle what it was not ; so Evolutionists seem to be trying to make out of Philosophical Anatomy what it does not really countenance.

The result in this case being as unprofitable as the Quiddities of the Schoolmen, we may, perhaps not unjustly, apply to the followers of Darwin the Poet's censure of the 'Sons of Aristotle : '

> ' They stand
> Locked up together hand in hand.
> Every one leads as he is led:
> The same bare path they tread,
> And dance like fairies a fantastic round,
> But neither change their motion, nor their ground.'

*

LECTURE II.

PROFESSOR HAECKEL'S SCHEME OF THE LINE OF MAN'S DESCENT FROM LOWER ANIMALS.

SUMMARY.

Mr. Darwin's recognition of the excellence of Professor Haeckel's exposition of the doctrine of Evolution, p. 36. Haeckel's assumption of the truth of the doctrine of Evolution and his estimate of the value of its influence in promoting the progress of Philosophy and Science, p. 36. His denial of any limit to natural knowledge, p. 37. *Anthropogenesis* treated of under the two distinct heads of *Ontogenesis*, or the embryonic development of the individual, and *Phylogenesis*, or lineal descent of the race by evolution from the most simple organisms, p. 38. Ontogenesis viewed as a short and rapid recapitulation of Phylogenesis, p. 39. The first living body alleged to have arisen as a mass of protoplasm, like a flake of white of egg, by spontaneous generation, p. 40. By development of a nucleus in a small mass of this protoplasm (with or without a cell-wall around it), a nucleated cell, such as an amœba, was formed, p. 40. By proliferation from this amœba, similar cells were produced, and these being aggregated together constituted a *Synamœbium*, p. 41. Which Synamœbium by an alleged '*differentiation*' of its component cells was evolved into a more complex organism, p. 41. Denial that any such process of 'differentiation' takes place in nature, p. 42. Such pretended '*differentiation*' constitutes the *false foundation* on which the superstructure of Evolution is in a great measure built, p. 43. True '*differentiation*,' wherein it consists, p. 43. The Ontogenesis of Protozoa a simple process of cell-development, p. 44. In the case of animals generally, the young individual originates from an egg, but the formation and development of the egg itself comprise the different stages of cell-life, p. 44. The *Blastoderma* composed of originally heterogeneous cells, not of homogeneous cells transmuted by an alleged process of '*differentiation*,' p. 45. The blastoderma in the various metamorphoses it undergoes towards the development of the embryo, considered by itself, resembles lower forms of organisation, p. 48. The embryonic development of *man* similar to that of other *mammifera*, but this fact is no proof that man was evolved from lower mammifera, p. 51. Nor is the fact of the resemblance of the embryo of mammifera to that of lower vertebrata any evidence of the evolution of a mammal from a frog or fish, p. 52. Equally little is the resemblance which may be traced between vertebrata and invertebrata any evidence that the lancelet, for example, was evolved from an ascidian mollusk, p. 57. Comparison of the plan of organisation of a lobster and that of a vertebrate animal, p. 53. The difference not so great between them, if we recognise that what is commonly considered the *ventral aspect* of the body of the lobster is really homologous with, or corresponds to, the *vertebral* or *dorsal aspect* of a vertebrate animal, p. 53. Unfounded nature of the alleged near relationship between an ascidian mollusk and the amphioxus or lancelet, by evolution of the latter from the former, p. 56. The structure in the tail of the ascidian larva, supposed to be the homologue of the *corda dorsalis* of the vertebrate embryo, denied to have any such signification, p. 56. The vertebrate ancestral forms through which Haeckel hypothetically alleges the human race to have been successively evolved were : the amphioxus or lancelet tribe of fishes ; lampreys ; fishes of the shark tribe ; fishes like the lepido-siren ; perennibranchiate batrachians, like the proteus and axolotl ; salamanders ; lizard-like creatures ; monotremata, or animals like the ornithorhyncus ; marsupialia, or animals like the kangaroo rat ; half apes, like the lemur ; tailed apes ; tail-less or men-apes, like the orang ; speechless-ape-men, p. 60. In this series of hypothetical animal forms transitional resemblances may be traced or assumed ; but this constitutes no evidence in favour of the doctrine of Evolution, p. 68. Evolution, in short, from beginning to end is an hypothesis *unverified* and *unverifiable*, p. 68.

LADIES and GENTLEMEN,—In the Lecture on Evolution, which I delivered in this theatre on the 29th of October last year, I examined only the more general and prominent points of the doctrine, as

pourtrayed in Darwin's ' Descent of Man.' Taking
now as a text, Professor Haeckel's recently
published work on 'ANTHROPOGENESIS,'[1] I propose on
the present occasion to give a connected view of the
alleged line of Man's Descent, tracing the ancestral
forms through which, according to Haeckel, the
human race was successively evolved.

Mr. Darwin, in his ' DESCENT OF MAN,' I may
premise, refers to Professor Haeckel's former work
on the ' HISTORY OF CREATION ACCORDING TO
NATURAL LAWS '[2] as affording a very complete
exposition of the doctrine of Evolution, and observes
that if the book had appeared earlier, he would
probably not have published his own. Haeckel's
present work comprises a course of lectures on the
Origin of Man, delivered to a popular audience at
Jena, in the university of which he is a Professor.

Assuming with all the zeal and earnestness of
conviction that the doctrine of Evolution is true,
and so self-evident that it cannot be justly gainsaid,
Professor Haeckel maintains that the recognition of
the descent of man throughout innumerable generations, from the lowest organisms, is the only means
calculated to guide us to a better comprehension of
the significance of the facts of *embryonic development*.
The better comprehension of the significance of the
facts of the embryonic development of man, thus

---

[1] *Anthropogenie —Entwickelungsgeschichte des Menschen*, von
Ernst Haeckel, Professor an der Universität Jena. Leipzig, 1874.
[2] *Natürliche Schöpfungsgeschichte*, von Dr. Ernst Haeckel, Professor an der Universität Jena. Berlin, 1870.

acquired, Haeckel, moreover, considers to be the light
which can alone illumine the path leading to real progress in Philosophy and Science.

That there is any limit to natural knowledge
such as is commonly assumed, Haeckel denies, and
bitterly censures   Professor Du Bois Raymond's
Essay on the subject, read at the Leipzig Meeting
of German Naturalists in 1873, protesting against
the Berlin Professor's '*Ignorabimus.*'

Suppose, argues Haeckel, that our *single-cell
amœban ancestors* of the Laurentian period could have
been told that their descendants of the Cambrian
period would become a *many-cell worm*, they would
not have believed it possible.   As little would these
worms have believed it possible that their descendants
would flourish as vertebrates like the *Amphioxus
tribe* and equally little would such *skull-less vertebrates*
have believed that descendants of theirs would be in
time developed into *animals with skulls.*   In like
manner, our Silurian *proto-fish ancestors* would never
have believed that any of their Devonian grandchildren would exist as *Amphibia* and their Triassic
great grandchildren as *Mammifera.*   So would those
mammifera have held it impossible that in the
Tertiary time, certain of their great great grandchildren would acquire *Human Form* and at the
same time become so highly gifted as to be qualified
to pluck the precious fruit of the tree of knowledge.
The unanimous answer from all would have been,
says Haeckel : 'IMMUTABIMUR ET IGNORABIMUS,'—

We shall never be changed—We shall never know it.'

It is this ' *Ignorabimus,*' continues Haeckel, with the unflinching confidence of a true Evolutionist— shall I say the dogmatism and intolerance also?— which Professor Du Bois Raymond would interpose as a barrier to the progressive development of science, and thus aid the 'Church Militant' in its crusade against freedom of thought and truth, reason and culture, development and progress;—a war in which Evolutionists, who are the *Soldiers of Truth*, can have no better ally than ANTHROPOGENESIS; for the real history of man's origin Haeckel considers to be the heavy artillery of the evolutionary army!

Such are the sentiments, Professor Haeckel frankly lets us know, with which he enters on his subject.

'ANTHROPOGENESIS' Professor Haeckel treats under the two distinct heads of *Ontogenesis* and *Phylogenesis* : Ontogenesis, or the embryonic development of the individual man by ordinary generation in the short period of 280 days ; and Phylogenesis, or man's alleged descent by Evolution, in the course of millions of years, from the lowest organisms, and in the last instance from apes.

Ontogenesis Evolutionists regard as a short and rapid recapitulation of Phylogenesis—the individual animal in the course of its development from the ovum to its complete form, passing rapidly through the most important of those metamorphoses, which its alleged successive progenitors in the course of innumerable generations slowly underwent by evolution, towards the form of its more immediate ancestors. Thus in the course of the development of a human being *in utero*, the embryo, at a certain stage somewhat resembles a fish, by and by a frog-like creature, and next a mammal—the mammalian form, be it observed, appearing first like that of the *Monotremata* or ornithorhyncus tribe the lowest of the class ; then like that of the *Marsupialia* or kangaroo tribe—afterwards like that of quadrupeds and apes, until, at last, the perfect human form is attained.

But before even the embryo appears, the ovum itself, in the various preliminary metamorphoses it undergoes in consequence of fecundation, presents to the eyes of Evolutionists recapitulations of the primæval and very simple forms which constituted, they allege, the beginnings of the *phylum* or line of man.

The facts and arguments adduced by Haeckel in support of these evolutionary views I proceed to examine.

What is called a *nucleated cell* is a microscopical corpuscle of *albuminous nature* in respect to substance ; and in respect to structure comprising two essentially distinct parts, viz. *protoplasm* and *nucleus*. These two parts may be enclosed within a membraneous vesicle—a third element of structure,—but this *cell-*

wall, though it has given name to the whole, is found not to be a constant or essential part of the structure of the body, for in certain cases an enclosing membraneous wall is never formed around the nucleated mass of protoplasm. But the nucleated cell, it is to be observed, does not represent the lowest stage of organic and living individuality. A more elementary phase of living existence is found in the *cytode*, a body which consists merely of a shapeless mass of protoplasm without a nucleus.

Of cytodes, nucleated cells and tissues developed from them, the bodies of animals are composed. But there are in Nature animal organisms of so simple a character as to consist of no more than a single cell, such as the *amœba*, or even no more than a cytode, such as a *protamœba*.

The amorphous living substance composing cytodes is assumed by Evolutionists to have originated in the first instance by spontaneous generation.

The elementary organism thus first presented itself as a *cytode*, and subsequently by the development of a *nucleus* in its substance and a *cell-wall* around it the phase of complete *nucleated cell* was attained.

From such hypothetical original single-cell animals like the *amœba* which, it is assumed, already existed in the early part of the primordial time, Professor Haeckel thinks it may be affirmed *à priori* that similar cells were produced by proliferation— that these new amœbæ formed by aggregation together a community to which he gives the name of *Synamœbium*; and that, though originally homogeneous in composition, structure, and endowments, they came at last by a process of 'differentiation' to form a many-cell organism possessed of differently endowed organs which all worked together for a common end.

To prove the validity of these phylogenetic presuppositions, Haeckel appeals to facts which ontogenesis or embryonic development, as he alleges, actually brings before our eyes at the present day, such as the formation of the *blastoderma*. This membrane is composed of cells into which the yolk of the egg is resolved in the process of division and subdivision which takes place after fecundation, which cells, though at first *homogeneous*, as he alleges, subsequently *differentiate* themselves, so as to form the various organs of the embryo. And thus, argues Haeckel, is his biogenetic fundamental law confirmed, viz., *that the development of the individual is a short and rapid recapitulation of that of the line.*

This argument seems to run in a circle, inasmuch as it assumes that cells having been spontaneously generated and having multiplied by proliferation, then set to work to 'differentiate' themselves, so as to become transformed into more complex animal bodies; whilst to prove this assumption another is made, viz., that the cells composing the blastoderma, though at first homogeneous, afterwards 'differentiate' themselves to form the more complex organs of the

embryo. Letting the fallacy of the argument pass,
I would comment on the root of it, viz., on the doctrine
of ' *differentiation*,' according to which homogeneous
cells may be developed into this or that, or indeed
any structure ; a doctrine which Haeckel and others
advocate as a fundamental part of the theory of
Evolution. Like as men in a community, says Haeckel,
occupy themselves each in a different employment
to the advantage of the whole, so the cells which
had become aggregated together to form a SYN-
AMŒBIUM, though originally homogeneous both in
substance and endowments, subdivided their labour
and engaged spontaneously in different modes of
action ; some metamorphosing themselves in one
way, some in another, and, at the same time, acquir-
ing new endowments so as to be qualified for the
performance of new functions.

This comparison would seem to imply that these
primæval cells worked with both purpose and will,
in order to acquire new forms and new endowments,
thus promoting their own evolution.   Our knowledge
of cells, however, does not warrant any such view of
their powers.   We see only that cells run through
the different phases of development, growth, and
metamorphoses proper to them, each kind according
to its own destiny, and that they thus unconsciously
fulfil the purposes of their life.

Granting any number of generations through
which cells may be supposed to have descended,

the Evolution argument, I contend, is not thereby strengthened.

When, by virtue of their original intrinsic qualities, differently composed and endowed cells are, under the appropriate conditions, developed into tissues, for example, each kind in its own proper way, we may truly call the process one of ' *differentiation*;' but there is certainly no such process, if by the expression is meant the development of different kinds of tissues from one and the same kind of cells by their own desires and efforts. The reality of such a process is, in fact, the very thing to be contested, as the *false foundation* on which the superstructure of Evolution has been, in a great measure, built.

Though resembling each other in their general characters, organic cells, let me repeat, are of various kinds—each kind possessing its own peculiar endowments—physical and vital—and its own mode of further development into tissues and organs. One kind of cells cannot give origin to another kind with different endowments, nor be developed by the chance of circumstances into this or that kind of structure. Cells react vitally under varying conditions, but each only in its own way. If the conditions be unusual, the cell still reacts after its own manner, according to its inherent endowments, so far as the unusual conditions permit. The cell may indeed become morbid, languish, and die under the

influence of unfavourable conditions, but it never
reacts so as to be converted into something of a
fundamentally different nature.

No such process as absolute '*adaptation*,' like
what Haeckel describes, can thus be admitted any
more than '*differentiation*.'

The elements in respect to shape composing the
various tissues of the animal body are formed out of
the substance of cells, their nuclei and intercellular
substance. And though thus similar in the *mechanism* of their development, tissues differ from each
other, not only in *shape* but also in *endowments*—
physical and vital—because they are developed each
tissue from a different kind of cells. In short, let
me reiterate : different cells, though they may
resemble each other in external aspect and general
points of structure, are potentially different in their
internal qualities. No transmutation, therefore, can
under any change of circumstances, take place of
one kind of cell into the tissue which is the proper
product of the development of another kind of cell.

The ontogenesis of Protozoa, we have seen, is
literally nothing more than an example of simple cell
development. In the case of animals generally, the
individual originates from an ovum or egg. The
first formation of the ovum, however, as well as its
development into the young animal, may be viewed
as merely different phases of the same process of
cell-life. Professor Haeckel regards the ovum itself,

as having the significance of a cell; the yolk repre-
senting the protoplasm, the germinal vesicle, the
nucleus, and the vitellary membrane, the cell-wall.
Gigantic a cell, I may remark, as the yolk of the
common hen's egg must, in that case, be looked on,
still more gigantic a cell must be the yolk of the
ostrich's egg, and more gigantic still must have been
the yolk of the eggs of the extinct birds of Mada-
gascar and New Zealand. But such a view of the
character of the ovum ought not to be entertained.
An ovum, on the contrary, must be considered as a
much more complex and very highly specialised
organism.

The germinal vesicle is not a nucleus with a
nucleolus, but is itself a nucleated cell. Through
the combined influence of it and the fecundating
cells, the yolk is constituted a special blastema, out
of which the blastoderma, or germinal membrane—
the structure from which the embryo originates—is
developed by a remarkable process of cell-formation.
The cells composing the blastoderma are, in their
endowments, potential as well as actual, of different
kinds, though formed in common from the yolk.
But the yolk, it is to be remarked, is a substance of
heterogeneous composition, and affords materials for
the development of the different kinds of cells, just
as a fluid containing different salts in solution yields,
on evaporation, different kinds of crystals. Hence,
in the development of the embryo, the various
tissues and organs have their origin in different

kinds of cells, and not in homogeneous cells by any alleged process of *differentiation* or *transmutation*.

In man and the mammifera, the ovum as it exists in the ovary or egg-bed, though so very minute in size as to be no more than visible as a speck to the naked eye, comprises all the essential parts of an egg, viz. the germinal vesicle, yolk, and yolk membrane. Professor Ernst von Baer, who first discovered the ovarian ovum of man and the Mammalia, mistook it for the homologue of the germinal vesicle of the bird's egg—which had been discovered a short time before by Professor Purkinje; but the discovery of a real germinal vesicle in the human and mammalian ovum by Valentin in Germany, Coste in France, and myself in this country, demonstrated the true nature and significance of that body. In the bird the ovum leaves the egg-bed as the full-sized yolk; whilst by the superaddition of the white and shell around the yolk, which takes place in the oviduct, the whole egg of the bird when laid, is, as is well known, larger still.

The first observable changes which take place in an ovum as the immediate consequence of fecundation are: the disappearance of the germinal vesicle, and the cleavage or division and subdivision of the yolk, resulting in the resolution of its substance into the different kinds of cells composing the blastoderma.

In the ovum after fecundation, when the germinal vesicle has disappeared, Haeckel thinks is to be recognised the ontogenetic representative of the spontaneously generated simple ·cytode, which he hypothetically assumes as the first ancestor of man, as well as of all other animals. To say nothing of the Evolution doctrine, in the interest of which Haeckel thus expresses himself, but taking his comparison of the newly fecundated ovum to a simple cytode as involving a question merely of homology, the view appears to me an exaggerated piece of transcendentalism. And not less extreme in its transcendentalism is the idea that the yolk of the ovum, after it has been converted into blasto-dermic cells by division and subdivision, is the ontogenetic recapitulation of a *Synamœbium*. The globular blastoderma of the mammiferous ovum, for example—whether we view it in respect to its forma-tion, or in respect to its wondrous though well-known and verifiable potentialities when formed—I ·hold to be a very different thing from an aggregation of simple homogeneous cells which can never give origin to anything else than cells like themselves.

When the ovum is minute in size, as in man and the mammifera for example, the yolk is, after fecunda-tion, all resolved into the cells forming the blasto-derma; but when the yolk is large, as in birds, it is in part only resolved into blastodermic cells,—this partial resolution into cells taking place in the region of the cicatricula, or spot where the germinal vesicle was imbedded. The explanation of the difference in size between the ovum of a bird and that of a

mammal is that the former, not having to be hatched
in a womb, contains all the material necessary for the
development of the chick, whereas the latter contains
sufficient materials only for the formation of the first
traces of the embyro ;—the materials for the further
development of the embryo and fœtus being derived,
within the mother's womb, from her blood through
the medium of the placenta.

To return to the ovum in the stage of globular blastoderma. In some animals, chiefly invertebrate, though also in the lancelet, and probably
in some higher vertebrata as well, the surface is
beset with vibratile cilia. This *planula-larva* form
Haeckel considers to be, ontogenetically, a recapitulation of a hypothetical many-cell protozoon which
he names PLAN.EA, and which he supposes first made
its appearance in the primordial time. In the next
stage of development the blastoderma becomes
separated into two distinct strata of cells, named the
primary germinal layers. The cells composing these
layers are different from each other in general aspect.
That they are different also in endowments and potentialities as regards future development, we shall see.
With the separation of the blastoderma into the two
germinal layers, respectively named :—the outer,
EXODERM (otherwise *animal* or *serous* layer), and
the inner, ENTODERM (otherwise *vegetative* or *mucous*
layer), an important step towards the fundamental
construction of the embyro is made.
    The ovum of man and the mammifera in this

stage of development, Haeckel compares to the
gastrula larva of many of the lowest animals—the
form value of which is that of a sac with a single
opening—the wall consisting of two layers—an
exoderm covered with vibratile cilia, and an *entoderm*
lining the cavity, while the single opening is the
protostoma or primitive mouth. The lowest animals,
such as sponges and the simplest polyps, continue
throughout life in this two-leaved form ; but the
ovum of man and the mammifera rapidly passes
through it towards the higher stages of development
proper to them.

To the correspondence in form value between the
gastrula larva and the ovum of the higher animals
in the stage of development under notice, Haeckel
applies his biogenetic fundamental law, thus :—Man,
and all other animals which pass through a stage
of development in which their body consists of an
exoderm and an *entoderm*, must have descended from
a primæval stem-form, the whole body of which con-
sisted throughout life of only two different strata of
cells or germinal layers, similar to the lowest zoophytes of the present day. We are, therefore, he
continues, justified in admitting that in early times
such a common stem-form did exist ; and to this
hypothetical being, which he considers full of
significance, Haeckel gives the name of GASTRÆA.

In all animals, except the zoophytes referred to,
the two primary germinal layers of the blastoderma
subdivide in the further course of development into

E

secondary layers, which constitute the foundation of the different systems of organs of the embryo.

The cells composing each of these secondary layers are at first, says Haeckel, quite similar, but they soon become *differentiated.* That the cells of one layer differ from those of another is evident; but that they were originally homogeneous not only in form but also in endowments and potentialities, and that they spontaneously *differentiated* themselves, is the very thing to be contested, as I have before shown. True *differentiation* consists in this :—Out of a collection of originally heterogeneous cells, one kind is metamorphosed into one structure,—another kind into another, in the further course of development, by virtue of their own inherent powers, just as, to repeat the illustration, different kinds of crystals are deposited from a mixture of the solutions of different kinds of salts.

The secondary layers of which the blastoderma comes to consist are found not only in the ova of man, the mammifera and other vertebrata, but also in that of mollusca, arthropoda, echinodermata, the higher worms and higher zoophytes—a fact in comparative ontogenesis which Haeckel interprets as being of the highest phylogenetic significance. To another mind, however, the fact only shows what has been long recognised, that all forms are constructed according to a common plan.

The most important phenomena of general sig-

nificance, according to Haeckel, observable in tracing the ontogenesis of man must be recognised in
the fact that the development of his body from the
very commencement takes place in the same manner
that the development of the body of other mammifera
does, and that all the peculiarities of individual development which distinguish the mammifera generally
from other vertebrata are to be found also in man.
This I admit, but I cannot admit the validity of the
inference which Haeckel draws from it in support of
the doctrine of Evolution.

.The development of the embryo of certain mammiferous animals—a rabbit or dog, for example—has
been well made out ; and in a few instances the
opportunity has presented itself of examining even
the human embryo, thrown off by miscarriage, at a
very early stage of its development ; and enough
has been thereby ascertained to show that, in main
features, the early human embryo does not differ
much externally from the correspondingly early
embryo of other mammiferous animals. But not-
withstanding that the ova of man and mammiferous
animals thus resemble each other, the ova of no two
kinds are exactly alike even in outward appearance ;
whilst potentially they differ from each other just as
the different kinds of animals do which spring from
them. Though the mechanism of the process of development into the embryo be the same, the resulting
animal is different. By virtue of their own inherent
vital endowments and potentialities, the constituent

cells of the blastoderma of any given kind of animal undergo their own peculiar metamorphoses, and, as development proceeds, the new being, built up with the structures into which they are fashioned, acquires more and more the characters of the parent animal. This is, I would repeat, an example of *true differentiation*.

Haeckel and other advocates of the Evolution doctrine, dwell upon the resemblance between the early embryo of a man and the early embryo of a dog. But here, as everywhere else throughout their arguments, they are led away by a superficial resemblance in outward appearance, and pass over unnoticed the internal qualities and hidden power, by which the two somewhat externally similar aggregations of cells are at last unerringly developed —the one into a *man*, and the other into a *dog*.

As to the resemblance of the human embryo in common with that of other mammiferous animals to the embryo of the lower vertebrata, it is to be observed that as the plan of development in all is similar, so at every stage there are indications of a similarity of structure ; but indications only.

Then again, as to a resemblance between the lowest vertebrata and the invertebrata, there is no doubt that here also, indications of similar organisation may be detected. In the ovum of arthropods— a *crustacean* like the lobster—for example, it is found

that the primitive trace of the embryo and the central
organ of the nervous system on the one hand, and
the blastodermic vesicle on the other, do not differ
in their topographical relationship to each other from
the corresponding parts in the ovum of the vertebrate
animal. Putting out of view now the vertebral
column and more complete development of the central
organ of the nervous system of the vertebrate animal,
it may be said that the subsequent difference between
the vertebrate and arthropod is that the aspect of
the body where the primitive trace and the central
organ of the nervous system appeared, becomes in
the vertebrate animal what is known as the dorsal
aspect or back of its body, and in the arthropod what
is commonly called the ventral aspect of its body or
belly. Fundamentally, however, the dorsal aspect
or back of a vertebrate animal, is homologous with
the so-called ventral aspect or belly of a lobster or
insect; and we shall be correct in saying that the
real difference between vertebrate and arthropodous
animals is—not that the central organ of the nervous
system or spinal marrow is situated along the back
in the vertebrate, and along the belly in the arthropodous animal; but that in their natural prone
position, vertebrate animals have their vertebral
aspect or back directed upwards, whereas the natural
position of arthropodous animals is the reverse of
this, being that in which the vertebral aspect of their
body is downwards—the aspect in them commonly
called the back being in reality, as I have before

said, that which is homologous with the belly of vertebrate animals.

To suit the inverted position of arthropodous animals, the mouth has been placed at what corre-sponds to the back of the head in vertebrate animals, and in order to permit the necessary communication between it and the stomach, which retains its relative position, the æsophagus or gullet passes through the æsophageal nervous collar, which may be viewed as the medulla oblongata of vertebrate animals split into lateral halves by a longitudinal fissure—a mere mechanical adaptation like the passage of a tendon through a split in another—without involving any real fundamental difference of structure or function.

The central organ of the nervous system in arthropoda—the so-called *ventral chain* of *ganglions*, but really and absolutely the homologue of the *spinal cord* of vertebrata, was shown in 1832 by my much respected colleague, the late Dr. Grant, to be com-posed of two columns—the one *ganglionic*, the other *non-ganglionic*. But in accordance with the mistaken view as to the real dorsal aspect of the lobster or insect's body, it was supposed that the relative position of the two columns—the *ganglionic* or sensitive 'below,' and the *non-ganglionic* or motor 'above'— was the reverse of what it is in vertebrate animals. From my demonstration, however, of the true dorsal aspect of arthropoda, it is evident that the relative position of the two columns *corresponds exactly* with that of the two columns of the spinal cord of vertebrata.

According to my demonstration of the true ventral aspect also, of arthropods, it is evident that the so-called 'dorsal' vessel is not dorsal, but really thoraco-ventral, and corresponds with the *heart* of vertebrate animals not only in function, but also in relative position—is, in fact, homologous with it.

In both the vertebrate and arthropodous animal, the trunk of the body is a chain of segments—in the former represented by the bodies of the vertebræ, in the latter by the rings of the external skeleton. But, even in arthropods, we discover evidences of an internal segmentation in the chain of ganglia forming the real posterior or sensitive column of their spinal cord.

Though the articulated extremities of arthropods are productions of their external skeleton, and as such may be compared to productions of the external or dermo-skeleton of vertebrate animals, like the dorsal, caudal, and anal fins of fishes, the same view cannot, perhaps, be taken of the wings of insects. It is worthy of observation that the wings of insects correspond to the limbs of vertebrate animals in relative position and in number, being turned to-wards the real ventral aspect of the body and being never more in number than two pairs,—sometimes there is only one pair ; and as there are vertebrate animals destitute of limbs altogether, so there are *wingless* insects.

The example of a transition of structure between the invertebrata and vertebrata just adduced is not, perhaps, such as would be recognised by Haeckel, seeing that, in his opinion, arthropods have but a remote phylogenetic relationship with the vertebrata. It is in my opinion more indisputable, however, than the notorious alleged homology between the embryo of the ascidian mollusk and that of the vertebrate amphioxus or lancelet, on which Darwin and Haeckel rely with so much confidence as an example of the connecting link between the invertebrata and vertebrata.

It is difficult to recognise in the chord-like structure as described by Kowalevsky in the transitory tail of the ascidian larva a homologue of the chorda dorsalis of the vertebrata. If the single ganglion of the ascidian correspond, as is supposed, to the so-called supra-æsophageal ganglion of worms, the side on which this single ganglion is situated must in that case be really the *ventral* (as I have shown) ; and consequently the body supposed to be a dorsal chord does not bear the proper relative position which in the vertebrata the *chorda dorsalis* bears to the spinal cord. In fact, Kowalevsky's cord has not, in my opinion, the significance of a *chorda dorsalis* at all.

This disallowance of the alleged close relationship in their ontogenesis between the ascidian and the lancelet need not of itself, however, cause any break in the phylogenetic chain, supposing for a moment such a chain existed, for, as I have shown, we have in the arthropods a very satisfactory transitional link between invertebrata and vertebrata; whilst from arthropods to mollusks, and downwards among the other subkingdoms of the invertebrata, quite as good transitional links may be also traced.

Though I thus most freely admit a transitional affinity between invertebrate and vertebrate animals, I repeat that I cannot see in it any corroborative evidence of a descent by evolution of the latter from the former. And as little can I see in the transitional affinities among invertebrate animals any evidence of evolution of the higher of them from the lower.

In his scheme of evolution Haeckel recognises seven types of animals, viz., Protozoa, Zoophytes, Worms, Mollusca, Echinodermata, Arthropoda, and Vertebrata.

From a branch of Protozoa he represents his hypothetical GASTRÆA to have been evolved. This ancient being, of which the *gastrula larva* is its recapitulation in characters of remarkable identity in the ontogenesis of the most different animals of the present day, must, Haeckel thinks, have existed in the Laurentian period of the Archolithic time.

From the 'Gastræa' two different lines of the animal kingdom, according to Haeckel, were

evolved :—in one direction the lower groups of
Zoophytes, and in another Worms. It is among the
Worms that the stem of Echinodermata, Mollusca,
Arthropoda, and Vertebrata is to be found ; but in
Haeckel's opinion, there is no relationship except
this remote one of common descent from Worms
between the Vertebrata and the three other subkingdoms mentioned. The vertebrata, he considers
with Mr. Darwin, as we have seen, were evolved
directly from ascidian worms. So far as regards the
study of Anthropogenesis, therefore, both ontogenetic
and phylogenetic, by far the largest part of the animal
kingdom may be excluded ; the only stems, in fact,
which bear upon ANTHROPOGENESIS being *Protozoa*,
Worms, and *Vertebrata*.

In entering on the consideration of the Phylogenetic branch of Anthropogenesis, a brief view of
the geological periods at which the various ancestral
forms of man were, according to Haeckel, evolved,
is here premised.

The PRIMORDIAL, or ARCHOZOIC or ARCHOLITHIC
time, comprising the *Laurentian*, *Cambrian*, and
Silurian periods, is supposed to have extended over
many millions of years—a duration longer than that
of all the subsequent times put together. It was in
the earliest period of this time that living beings
originated by spontaneous generation. From these
organisms, which were of the most simple nature,
sprang all the invertebrate ancestors of man,—from

these again, the lancelet and its congeners were evolved,—and from these lowest vertebrate forms, Selachian and Ganoid fishes which existed in the *Silurian* period, descended.

As yet all organised and living beings were inhabitants of the waters.

To the Primordial time succeeded the PALÆO-LITHIC, or PALÆOZOIC or PRIMARY, comprising the *Devonian, Carboniferous,* and *Permian* periods. The Palæolithic time was much shorter in duration than the *Archolithic,* but longer than the subsequent times. In the Carboniferous period existed *Amphibia,* the most ancient land and air-breathing vertebrates.

In the MESOLITHIC time, comprising the *Triassic, Jurassic,* and *Chalk* periods, osseous fishes first appeared, but this time was, in an especial degree, that of reptiles—the gigantic dragons and so-called flying lizards. During this time also birds began to appear as offshoots from lizards, of which the Archeopteryx is an example—a bird with a lizard's tail. Mammals also appeared in the form of marsupialia.

In the CAENOLITHIC or TERTIARY time, comprising the *Eocene, Miocene,* and *Pliocene* periods, which was short, mammiferous animals lived in full development, and the transition from apes to man commenced.

The next, or QUATERNARY time may, Haeckel suggests, be designated the ANTHROPOLITHIC, as it was the time when men, endowed with the faculty of speech, were evolved from speechless ape-men, and when the full development of the various human races took place. This ANTHROPOLITHIC time comprised the *Glacial, Post-Glacial,* and *Historic* periods.

Under the head of Ontogenesis, I have anticipated all that need be said relating to the Invertebrate stages of the alleged phylogenesis of man ; I shall, therefore here take up the line with the Vertebrate stages, and then conclude with a retrospective summary of the whole.

The series of man's vertebrate ancestors begins, according to Haeckel, with a hypothetical animal which lived during the primordial time, and of the characters of which we have presented to us a distant idea in the still living *amphioxus* or *lancelet.*

The lancelet tribe just mentioned forms Haeckel's ninth stage in the phylogenesis of man. His tenth stage comprises *cyclostomatous* fishes, or fishes of the *lamprey* tribe, which he names *Monorrhina* or single-nosed, as they have only one nasal cavity—a character in which they differ from true fishes.

The eleventh ancestral stage of man was represented by *proto-fishes.* These, Haeckel considers, were evolved from monorrhina by the division of the single nostril into two lateral cavities, and by the acquisition of branchial arches and a jaw skeleton,— of a swimming bladder and two pairs of limbs in the shape of pectoral and abdominal fins—besides other characters. These proto-fishes, Haeckel thinks, resembled the lowest *squali* (or fishes of the shark tribe) of the present day, and already lived in the

Silurian period, as is indicated by the fossil remains
of teeth and fin spines.

Our twelfth ancestral stage, according to Haeckel,
was represented by animals which probably possessed
a distant resemblance to the still living *lepido-siren*
tribe of fishes. They were evolved from proto-fishes
by metamorphosis of the swimming bladder into a
lung, and conversion of the nasal cavities into air-
passages opening into the mouth.

With this stage began the series of man's ancestors breathing by lungs. A branch of Proto-fishes
or Selachii during the Devonian period, Haeckel
thinks, made the first successful attempt—*the first
successful attempt*, to repeat the expression—to live
on land and breathe air; whereby the swimming
bladder became a lung with a corresponding change
in the structure of the heart.

From the amphibious fishes of the twelfth stage
just described, the *perenni-branchiate batrachians*,
forming the thirteenth stage, were evolved. The
extinct unknown form from which man descended,
Haeckel designates by the name of AMPHIBIUM. It
retained its gills throughout life like the still living
proteus and axolotl; and its evolution from the
amphibious *lepido-siren* took place by metamorphosis
of the paddle-like fins into legs with five-toed feet,
and by a higher 'differentiation' of various organs,
especially the vertebral column. This perenni-
branchiate *amphibium* existed about the middle of
the Palæolithic or Primary time—perhaps before

the Carboniferous period; for fossil Amphibia are already found in coals.

Following our amphibian forefathers, as Haeckel calls them, which retained their gills permanently, there appeared other Amphibia which by metamorphosis at a later period of their life lost their gills, but retained their tail. These hypothetical ancestors of man forming the fourteenth stage of his line, Haeckel thinks, were similar to the *salamanders* and *tritons* of the present day. They originated from the gilled amphibia by accustoming themselves—*accustoming themselves*, let me repeat the expression—at an advanced period of their life to breathe only by their lungs! Probably, they already lived in the second half of the Primary time—during the Permian or perhaps the Carboniferous period. The proof that such animals once existed is, in Haeckel's opinion, that tailed amphibia form a necessary middle link between the preceding and the following stage. Let me repeat this exquisite specimen of easy assumption, and evolutionary argument : ' *The proof that such animals once existed is, that tailed amphibia form a necessary middle link between the preceding and the following stage !* '

The *fifteenth* phylogenetic stage was represented by unknown lizard-like forms, which Haeckel names *Protamnia*, as being the forerunners of those animals the embyro of which, in the course of its development *in ovo*, is enclosed by the membrane called. *Amnion*. Their advent dates probably from the beginning of the Mesolithic or Secondary time—perhaps already from the Permian period of the Primary time.

The *Protamnia* were the common stem, from which were evolved in two diverging branches, Reptiles and Birds on the one hand, and Promammalia on the other. With the *Promammalia* only we have to do as ancestors of man.

Among our forefathers, from the *sixteenth* to the *twenty-second* stage, we are, says Haeckel, more at home, as they all belonged to the great and well-known class of Mammalia. The now long extinct and unknown common stem-form of all Mammalia, which Haeckel designates by the name of Promammal, stood nearest the Ornithorhyncus and Echidna of all now living animals.

Promammalia were evolved from Protamnia probably in the Triassic period of the Mesolithic or Secondary time, by 'increased development of the internal organisation,' by the formation of a lacteal gland to supply milk for the nourishment of the young, and by the conversion of epidermic scales into hairs.

Marsupialia, or animals of the Opossum and Kangaroo, tribe, constitute the immediate transition between the Monotremata and Placental Mammalia as well in an anatomical as in an ontogenetical and phylogenetical respect. Among the marsupialia, therefore, ancestors of man must also have been found. They originated from the Promammalian

Monotreme, by the partitioning off of the cloaca into rectum and urogenital sinus, by development of a nipple to the lacteal gland, and by the partial retrogression of the clavicle.

The most ancient of the marsupials lived in the Jurassic period of the Mesolithic time—perhaps already in the Triassic period ; and, during the Chalk period, Placental Mammals began to be evolved from them.

The immediate stem-form of *true apes* and through them of *Man* also comprised *Prosimiæ* or half-apes, which originated probably in the beginning of the Caenolithic or Tertiary time, from unknown Marsupials related to the marsupial or Kangaroo Rats, by the formation of a placental connection between the mother and fœtus, by the loss of the marsupium or pouch, and the marsupial bones supporting it, and by the greater development of the *corpus callosum* of the brain. These our half-ape ancestors probably possessed only a remote external resemblance to the now existing short-footed half apes, such as the lemur.

Of the *true apes* which were evolved from the *half apes*, the CATARHINE or thin-nosed tribe of the old world alone possesses a near blood relationship to man. Our extinct ancient forefathers of this group probably resembled the nosed ape of the present day, though still covered with hair, and possessing a long tail. They made their appearance about the Eocæne period of the Tertiary time.

Of all still living apes, the nearest to man are the great tail-less thin-nosed apes, the Orang and Gibbon of Asia, and the Gorilla and Chimpanzee of Africa. These anthropoid apes probably originated in the Miocæne period of the Tertiary time, as did the unknown direct ancestors of man now extinct.

Although the preceding *man-like apes* already stood so near to man proper that the admission of a transition stage is scarcely needed to complete the line, we may, nevertheless, Haeckel says, consider as such, speechless *Ape-men*. These 'FATHERS OF MAN' probably lived towards the end of the tertiary time and were evolved from *man-like apes* by complete habituation to the upright posture, and the correspondingly greater differentiation of the pectoral from the pelvic extremities, the anterior hand becoming in them the human hand, and the posterior hand a walking foot.

TRUE MEN were evolved from *Ape-men* of the preceding stage by the gradual development of their vocal sounds into connected and articulate speech. The acquisition of this faculty being naturally accompanied hand in hand by the development of other organs, and by the higher 'differentiation' of the larynx and of the brain.

This transition from *speechless* APE-MEN to the *true* or *speaking* MEN first took place, probably, in the beginning of the quaternary, or, as Haeckel proposes to call it, the Anthropolithic time, that is the glacial

F

period, though, perhaps, it may have already taken
place in the Pliocæne period of the Tertiary time.

Such is Professor Haeckel's phylogenetic scheme.
After the arguments adduced in the preceding part
of this lecture against phylogenesis in general, it
would be supererogatory to make any further comment on it here. I may, however, in addition to the
remarks on the evolution of language made in my
first lecture (p. 27), apply to what Haeckel says of
the development of speech, a word from Professor
Max Müller. That eminent philologist strongly
dissents from the idea that language could ever have
had any connection with the alleged evolution of
man from apes, and maintains that, at the outset,
the possession of the faculty of speech by man constituted a definite and distinct line separating him
from the lower animals. ' Language, and what is
implied by language,' Müller considers to be ' the
specific difference between man and beast.'

' All the materials of our knowledge,' continues
Müller, ' we share with animals. Like them we
begin with sensuous impressions, and then, *like ourselves* and *like ourselves only*, proceed to the GENERAL,
the IDEAL, the ETERNAL. In many things, indeed,
we are like the beasts of the field ; but *like ourselves*
and *like ourselves* only, we can rise superior to our
bestial self, and strive after what is UNSELFISH, GOOD,
and GODLIKE.'

To take now, in conclusion, a summary retrospect of Haeckel's scheme of the *Phylogenesis* or line of descent of Man :—

MAN as he now is was originally evolved from hypothetical speechless *Ape-men*; these ape-men, again, were evolved from hypothetical *Men-apes without tails*, like the Orang ; these men-apes, from hypothetical *Apes with tails*, like the nosed apes ; these tailed apes, from hypothetical *Half apes*, like the *lemur*; these half apes, from hypothetical *Marsupial animals*, like the *Kangaroo rat*; these marsupialia, from hypothetical *Monotremata*, like the *ornithorhynchus*, but without the duck's bill ; these monotremata, from hypothetical *Lizard-like* creatures of which no living resemblance is known ; these lizard-like creatures, from hypothetical water *Newts* or *Salamanders*; these salamanders, from hypothetical *Perenni-branchiate batrachians*, like the *proteus* or *axolotl* ; these perenni-branchiates, from hypothetical *fishes*, like the *Lepido-siren*; these double breathing fishes, from hypothetical fishes of the *Shark* tribe ; these proto-fishes, from hypothetical *Lampreys*; these cyclostomata, from hypothetical *Amphioxi* or *Lancelets*, which lowest of vertebrate animals again were evolved from a hypothetical form of *Ascidian mollusk* or *worm*, low in the scale of invertebrate animals ; these ascidian worms from hypothetical *soft or cavitary worms*; these from hypothetical *solid worms*; these from hypothetical *Gastræada* ; these from hypothetical *Planæada* ; these from hypothetical *Synamoebia*, consisting of a community of homogeneous cells ; these, protozoa from hypothetical *single cell animals* ; and these, lastly, from hypothetical *spontaneously generated cytodes.*

Though this, it must be confessed, is a lame and impotent conclusion, still Haeckel's phylogenetic disquisitions · have one great scientific merit, that, namely, of leading to close and searching inquiries into the natural affinities of animals in succession from the lowest to the highest ; but, as I contend, without proving any transition by evolution or lineal descent. In thus rejecting the doctrine of Phylogenesis, I necessarily, at the same time, ignore the teachings as to the true efficient causes of Ontogenesis, which Haeckel most confidently claims for it.

We thus see that EVOLUTION, from beginning to end, is an *unverified* and *unverifiable hypothesis.* The scheme may, indeed, be entertained, as it has long been, more or less, as suggesting inquiries into the natural affinities of organised beings ; and in this respect I have just eulogised Haeckel's phylogenetic disquisitions. But when the doctrine is promulgated as a kind of *new revelation in Science,* and obtruded on us almost as an *article of faith* in a propagandist and intolerant spirit, we are roused to repel the attempted encroachment.

Our present advanced knowledge in Natural Science has not rendered the idea of Evolution a bit more probable than it was in former times. And it must be firmly denied that the conceit of Natural Selection by Survival of the Fittest has, in any degree, imparted to the theory more substantial body than it had before, or raised it to the scientific position which Darwin and his followers claim for it.

LONDON: PRINTED BY
SPOTTISWOODE AND CO., NEW-STREET SQUARE
AND PARLIAMENT STREET

.

www.ingramcontent.com/pod-product-compliance
Lightning Source LLC
Chambersburg PA
CBHW021954190326
41519CB00009B/1255